Where to Find Minibeasts

Minibeasts In the Soil

Sarah Ridley

A⁺

Smart Apple Media

Smart Apple Media
P.O. Box 3263, Mankato, Minnesota 56002

Printed in the United States

Published by arrangement with the
Watts Publishing Group Ltd, London.

Library of Congress Cataloging-in-Publication Data

Ridley, Sarah.
   In the soil / Sarah Ridley.
      p. cm. -- (Where to find minibeasts)
   Includes index.
   ISBN 978-1-59920-325-6 (hardcover)
   1. Soil invertebrates--Juvenile literature. I. Title.
   QL365.34.R525 2010
   592'.1757--dc22
                                    2008044906

Series editor: Sarah Peutrill
Art director: Jonathan Hair
Design: Jane Hawkins
Illustrations: John Alston

The measurements for the minibeasts in this book are typical sizes for the type of species shown in the photograph. Species within groups can vary in size enormously.

Picture credits and species guide:
front cover t: Boy looking through magnifying glass, Sonia Etchison/Shutterstock. **front cover b:** Millipede (*Narceus annularis*), Octavian Florentin Babusi/Shutterstock. **2:** Common Earthworms (*Lumbricus rubellus*) in soil, Phil Morley/Shutterstock. **3:** Centipede (*Haplophilus subterraneus*), Nigel Cattlin/FLPA Images. **6t:** Common Earthworm (*Lumbricus rubellus*) in soil, Derek Middleton/FLPA Images. **6b:** Ants (family *Formicidae*) at entrance to their nest, Ljupco Smokovski/Shutterstock. **7:** Red slug (*Arion rufus*) feeding, François Merlet/FLPA Images. **8t:** Nest mound of mining bee (*Halictus sp*), Nigel Cattlin/FLPA Images. **8b:** Mining Bee (*Colletes sp*) feeding, K. Smith/FLPA Images. **9t:** Mining Bee female (*Saropoda bimaculata*) digging nest, Martin Gerrard/NHPA. **9b:** Bumble Bee (*Bombus terretris*) nest, Paulo de Oliviera/OSF. **10t:** Nest of the Common Wasp (*Vespula vulgaris*), Stephen Dalton/NHPA. **10b:** Wasp (*Vespula vulgaris*) nest showing eggs & larvae, Scott Camazine/OSF. **11t:** Common Wasp (*Vespula vulgaris*) queen, Bob Gibbons/FLPA Images. **11b:** Cells of Common wasp (*Vespula vulgaris*) nest, Kurt Hahn/istockphoto. **12:** Casts of Common earthworm (*Lumbricus rubellus*), Robert Pickett/Papilio. **13t:** Common Earthworm (*Lumbricus rubellus*), Goga/Shutterstock. **13b:** Male Blackbird (*Turdus merula*) hunting, David Dohnal/Shutterstock. **14:** Vegetable garden, Darryl Sleath/Shutterstock. **15:** Making a wormery, Ray Moller/Franklin Watts. **16t:** Centipede (*Haplophilus Subterraneus*), B. Borrell Casals/FLPA Images/Corbis. **16b:** Centipede (*Haplophilus subterraneus*), Nigel Cattlin/FLPA Images. **17t:** Centipede (*Haplophilus subterraneus*) killing Common Earthworm (*Lumbricus rubellus*), Hecker/Sauer/Still Pictures. **17b:** Centipede (*Scolopendra sp*) with eggs, OSF. **18t:** Flat-backed Millipede (*Polydesmus angustus*), Richard Ford/Digital Wildlife. **18b:** Snake Millipede (*Proteroiulus fuscus*), Milos Luzanin/Shutterstock. **19:** Pill Millipede (*Glomeris marginata*), François Merlet/FLPA Images. **20:** Cockchafer (*Melolantha melolontha*) larva, David T. Grewcock/FLPA Images. **21t:** Sexton Beetle (*Nicrophorus humator*), Hecker/Sauer/Still Pictures. **21b:** Green Tiger Beetle (*Cicindela campestris*) larvae, Richard Becker/FLPA Images. **22:** Snail (*Otala punctata*) laying eggs, B. Borrell Casals/FLPA Images. **23t:** Yellow Ants (*Lasius flavus*) with pupae in nest galleries, Laurie Campbell/NHPA. **23b:** Meadow grasshopper (*Chorthippus parallelus*), Alexander Vasilyevich lmodeev/Shutterstock. **24t:** Magnifying glass, Maugli/Shutterstock. **24c:** Brush, Adam Borkowski/Shutterstock. **24b:** Boxes, Scott Rothstein/Shutterstock. **25t:** Pseudoscorpion (*Lasiochernes cretonatus*), David M Dennis/OSF. **25bl:** Girl, Ray Moller/Franklin Watts. **25br:** Hands, Shutterstock. **31:** Common Earthworms (*Lumbricus rubellus*) in soil, Phil Morley/Shutterstock.

Contents

Words in **bold** are in the glossary on pages 28–29.

 Your parent or guardian may want you to wear disposable gloves or gardening gloves before you poke around in the soil.

Under Your Feet

The ground under your feet is a **habitat** to many minibeasts, such as worms, millipedes, and centipedes. For many minibeasts, the soil is where they make their homes and find their food.

Worms can grow up to 12 inches (30 cm) long.

▲ An earthworm tunnels up through the soil.

This ant can grow up to ⅕ inch (0.4 cm) long.

▲ Ants at the entrance to their nest.

What Is a Minibeast?

Minibeast is the name given to thousands of small animals, from ants to millipedes and wasps to worms. Although many are **insects**, others are not. None of them have a **backbone**, so scientists give them the name "**invertebrate**."

Underground Homes

Many minibeasts build their homes underground. Even small minibeasts, such as ants, can move grains of soil to make tunnels and chambers as they build a nest.

Rotting Recyclers

Several minibeasts that live in soil break down dead plant material by eating it. They **recycle** it back into the soil, where it helps plants grow. Without these minibeasts, there would be piles of dead and **rotting** plant material in our open spaces.

TOP TIP!

Learn to use your eyes and you will see holes in the ground that lead to minibeast homes. Look for more boxes like this to help you become a good minibeast spotter.

▼ Slugs eat garden vegetables, wild mushrooms, and plants. They can often be seen traveling across soil in search of food.

The red slug can grow up to 7 inches (18 cm) long.

Watch Those Holes!

A sunny bank can become home to **solitary** bees. Look out for them popping in and out of their earth holes.

Bees All Alone

Many types of bees live alone in their own **burrow**, rather than in a **hive** or **colony**. Each bee digs out a small tunnel leading to an underground chamber. Inside her hole, the bee lays her eggs, which hatch into **larvae** and eventually grow into adult bees.

▲ If you find a hole like this against a wall or near a path, it may be a bee nest.

▼ A bee collects plant **pollen** and **nectar** to feed to the larvae.

This bee can grow up to ½ inch (1.4 cm) long. ⊢⊣

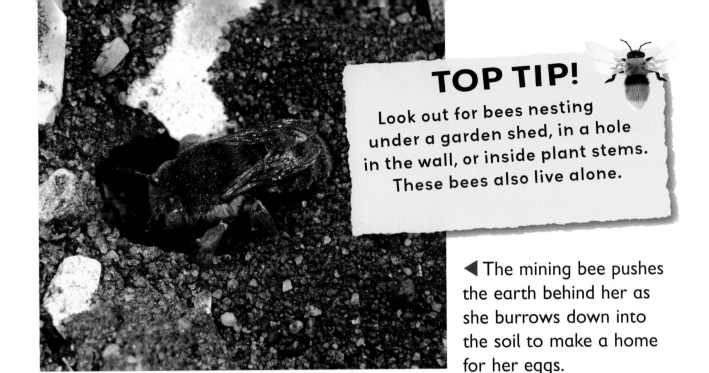

TOP TIP!

Look out for bees nesting under a garden shed, in a hole in the wall, or inside plant stems. These bees also live alone.

◀ The mining bee pushes the earth behind her as she burrows down into the soil to make a home for her eggs.

Hibernating Bumble Bees

Bumble bees live in small colonies. In the autumn, the colony dies off as the weather grows cold. Only the **queen** survives, burrowing down into the soil or under some leaves. There she drops into a deep sleep called **hibernation**. When she awakes, she crawls out, starts feeding, and then finds a good place to lay her eggs and start a new colony.

▶ This is an underground bumble bee colony.

Underground Wasps

Many wasps, like solitary bees, live alone in holes in the ground. Others live in nests containing thousands of wasps.

Getting Started

In the spring, a queen wasp searches for a good place to build a nest, such as an empty rabbit burrow. She builds a nest and lays some eggs. The eggs hatch into larvae that she feeds until they turn into worker wasps.

▲ A badger has damaged this wasps' nest, built inside a rabbit burrow.

Bigger and Bigger

The worker wasps build up the nest. The queen lays a large number of eggs. Other worker wasps feed the larvae hatching from the queen's eggs.

◀ This wasps' nest has been split open to show the eggs and larvae inside.

▶ Wasps build their nests from pieces of wood. Look out for wasps on wood such as garden furniture. They collect tiny pieces of wood, mix them with **saliva**, and use this mixture to build up the nest in layers.

TOP TIP!

If you see several wasps coming and going from the same spot, you have probably found a wasps' nest. Keep your distance! Wasps will defend their nest by stinging you. ⚠️

This wasp can grow up to ⅔ inch (1.7 cm) long.

Hexagonal Homes

A wasps' nest is made of six-sided (hexagonal) paper **cells**. Each cell contains one egg. The eggs hatch into hungry larvae that eventually grow into adult wasps.

◀ This close-up of the inside of a wasps' nest shows how the cells fit together.

11

Mud Swirls

Have you ever seen little swirls of mud on the lawn or at the park? These are worm casts, or worm waste. Worms leave them behind on the surface and then burrow back down into the soil.

Wiggling Worms

Worms spend their lives burrowing underground. They make tunnels by pushing through the soil and eating some of it as well. Any dead plant material in the soil becomes the worms' food. The waste passes through the worms and comes out as smooth, wet soil.

▲ A worm cast does not smell—it is just a pile of soil that has passed through a worm's body.

Head or Tail?

Most worms are a pale pink color. The head is the pointed end and the tail is the flatter end. If you can't tell, check the direction in which the worm is moving —the head will be at the front.

▲ Some people think a worm will survive if it is cut in half. It won't!

Worm Lovers

Lots of animals like to eat worms for food. Birds, hedgehogs, badgers, moles, shrews, and some minibeasts think worms are tasty.

► Worms are good food for birds and other animals.

Important Worms

▲ Under the ground, worms will have helped these plants to grow well by mixing up the soil, making air spaces, and making the soil **fertile**.

Watch how worms mix soil by making this simple worm farm.

You will need:
- 2 jugs • Soil • Sand
- A glass jar, tank, or plastic container
- 3 or 4 earthworms • A few leaves
- Black paper
- Scotch tape

What to do:

- Fill one jug with soil and another with sand. Fill the container with layers of sand and soil in turn.

- Add the earthworms and some leaves.

- Wrap a black paper cover around your container and tape it down.

- Keep the worm farm in a cool place.

- After a few days, take the black paper off to see what the worms have done to the layers.

- Remember to return the worms to where you found them, as they can eventually dry out and die.

Crazy Centipedes

This centipede can grow up to 1.4 inches (3.5 cm) long.

▲ A centipede has a pair of legs on each **segment**, or section, of its body.

Dig in the soil and you may see a long, pale minibeast with many legs. This is a centipede. Its name means "one hundred legs," but in fact some centipedes have fewer than 100 legs and others have more.

▼ Centipedes that live in soil can tie themselves in knots!

Dark and Damp

Centipedes are **nocturnal** and like to rest in dark, damp soil during the day. When the weather is hot and the sun dries the upper layers, centipedes burrow down to where the soil is still damp.

TOP TIP!

To search for centipedes, use a spade to turn over some soil in the garden. Ask an adult first! Some types of centipedes live under stones or in compost heaps, so try searching for them there as well.

Night Hunters

At night, centipedes become fierce, fast **predators**. With their flat, flexible bodies, they can chase another minibeast wherever it goes—even under rocks. They eat beetles, spiders, slugs, snails, worms—and each other!

▲ A centipede kills a worm by injecting it with poison.

Caring Parent

A mother centipede lays her eggs in the soil. The centipedes grow, and eventually have their own children. When the tiny centipedes hatch, the mother stays with them until they are old enough to survive alone.

◄ The mother centipede curls her body around the eggs while she waits for them to hatch.

Munching Millipedes

Millipedes push through the soil, or along the surface, searching for food. Their smooth, strong bodies help them to slide through the soil or **leaf litter**, pushed along by all those legs.

Slow Movers

Despite the name, which means "one thousand legs," no millipede has as many as that. Millipedes do have lots of legs though—two pairs on each segment of their bodies.

This flat-backed millipede can grow up to .9 inch (2.4 cm) long.

▲ The flat-backed millipede looks a bit like a centipede.

This snake millipede can grow up to 1.2 inches (3 cm) long.

▲ This is a snake millipede. A millipede moves slowly on its short legs.

Helpful Plant-Eaters

Millipedes do not need to be speedy because they eat dead plant material. They help us all by breaking down rotting or dead plants.

Defense

When a millipede feels danger, it curls up in a spiral. This makes it more difficult to attack. Some millipedes also squirt a foul-smelling liquid that keeps many animals from eating them.

TOP TIP!

Millipedes are nocturnal. Look for them resting in dark, damp places like soil, under stones, or under the bark of old trees.

▲ The pill millipede looks very like the pill woodlouse.

How to tell a millipede from a centipede

MILLIPEDE:	CENTIPEDE:
slow mover	fast mover
two pairs of legs per segment	one pair of legs per segment
rounded body	flat body
short **antennas**	long antennas

Beetles and Their Babies

There are thousands of different types of beetles worldwide. Some beetles live in water, others live in the desert, and some live in gardens.

Greedy Grubs

The female cockchafer beetle digs down into the soil to lay her eggs. When they hatch, the **grubs** stay underground for about two years. They feed on plant roots and can cause a lot of damage.

▶ Gardeners and farmers view cockchafer grubs as pests.

This cockchafer grub can grow up to 1.8 inches (4.5 cm) long.

TOP TIP!

Another place to find beetle grubs is under the bark of rotting trees. You may also see woodlice, centipedes, and millipedes.

Sexton Beetles

Sexton beetles make sure their grubs will have a good supply of food when they hatch. They dig out some of the soil underneath a dead animal, such as a mouse. The body sinks down until it is almost covered by soil. Then the beetles lay their eggs in the soil close by. When the grubs hatch, they feed on the dead animal.

This sexton beetle can grow up to 1.2 inches (3 cm) long.

▲ The adult sexton beetle also eats dead animals. It helps to speed up the rotting process.

▲ Look for small, round holes in sunny spots. Can you spot the tiger beetle grubs ready to pounce?

Terrible Tiger Grubs

The tiger beetle grub digs itself a pit and lurks inside. When a minibeast passes by, the grub lunges out to catch it or waits for it to fall into the pit.

These tiger grubs can grow up to 1.8 inches (4.5 cm) long.

Minibeast Eggs

Many minibeasts lay their eggs in the soil. Some build amazing homes for their eggs and larvae, a few protect their eggs until they hatch, and others just bury their eggs in the soil and leave.

Shiny White Balls

Snails and slugs lay their shiny, white eggs in damp soil or other places where the eggs will not dry out. Depending on how warm the soil is, the eggs hatch after a few weeks.

▶ This snail has pushed its body into the soil to lay eggs.

This snail's shell can grow up to 1.6 inches (4 cm) across.

Ant Eggs

Ants live in big groups. Their homes are called colonies. While some ants build under a stone, others go down a crack in the earth and start building. Ants dig out tunnels to connect the chambers where they care for their eggs and larvae.

▲ Ant eggs are white and look like small grains of rice.

This grasshopper can grow up to ¾ inch (2 cm) long.

▲ A grasshopper pushes the tail end of its body into the soil and lays eggs close together under the ground.

A Handful of Soil

To find some of the minibeasts that live in soil, follow the activity shown here.

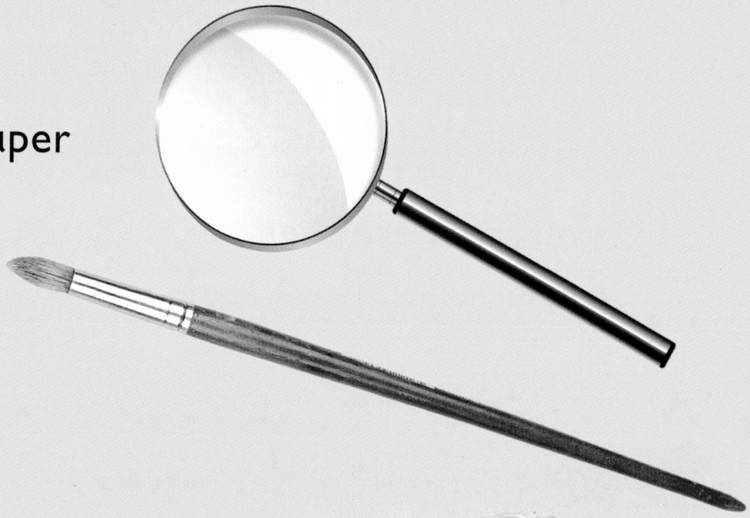

You will need:
- A large sheet of white paper
- Disposable gloves
- A handful of garden soil
- A fine paintbrush
- Collecting tubs
- A magnifying glass or microscope

What to do:
- Lay the sheet of paper on a table.
- Wearing the gloves, carefully grab a handful of soil.
- Carefully place the soil onto the white paper and spread it around.
- Using the paintbrush, slowly move the soil around to look for minibeasts.
- Carefully lift the minibeasts into a collecting tub.

- The magnifying glass will help you to see things more clearly.

- Use the identification guide on pages 26–27 to help you.

- Remember to return the soil and the minibeasts to where you found them.

▼ You could repeat this activity using a handful of leaf litter. Minibeasts that live in soil are often found in leaf litter also.

▲ A magnifying glass may help you to see a pseudoscorpion. This minibeast lives in leaf litter, soil, or compost.

This pseudoscorpion can grow up to ⅛ inch (0.3 cm) long.

⚠ Wash your hands after you have handled soil.

Identification Guide

Use this guide to help you identify the minibeasts that you find. They are listed in the order in which they are featured in the book. Some other common minibeasts you might come across are listed at the end. Because there are thousands of different minibeasts, you may need to use a field guide or the Internet as well.

Bee: Bees are flying insects. Some bees live in huge, complex nests or hives; others live in holes in the ground. All bees sting if they feel they are in danger, so don't touch!

Millipede: A millipede is long and slim and moves slowly. It has 2 pairs of legs on each body segment. There are several different types (see pages 18–19).

Wasp: A flying insect with two pairs of wings. Wasps build nests underground or in roof spaces. If wasps feel they are in danger they will sting, so don't touch!

Beetle: An insect with two pairs of wings. The front wings are a hard, shiny wing case that protects the soft back wings, which are used for flying.

Worm: A long, pale minibeast that spends most of its life tunneling underground.

Beetle grub: The young of a beetle, also called a larva. Some grubs live underground, others in rotting wood.

Centipede: A fast-moving, flat, long minibeast with one pair of legs on each of its body segments. It hunts other minibeasts.

Snail: The snail carries its shell on its soft body and is a **mollusk** like the slug.

Slug: A mollusk, the slug has a shiny, wet body and moves along on its one "foot."

Harvestman: A harvestman has eight legs and a body in one piece. It belongs to the arachnid family of animals, like the spider.

Ant: The ant is an insect and is related to wasps and bees. There are thousands of different types of ants worldwide. They build nests, also called colonies.

Mite: This tiny animal has eight legs and is an arachnid, like the spider.

Slug, snail, and ant eggs: Snail and slug eggs are shiny, damp, white balls. Ant eggs are small, oval, white or yellow eggs, similar in size to a grain of rice.

Springtail: This tiny insect is about the size of a pin head. There are many types, but they all have six legs, two antennas, and a special spring under their tail. This allows them to jump.

Grasshopper: An insect with strong back legs for jumping, as well as wings for flying.

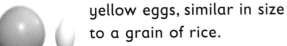

Spider: Spiders come in all shapes and sizes, but they all have eight legs and a body in two parts. They belong to the arachnid family and are not insects.

Pseudoscorpion: A tiny **arachnid**, from the same minibeast family as the spider; the pseudoscorpion has large pincers at the front of its body.

Silverfish: Silverfish have remained almost the same since the time of the dinosaurs. These small, gray insects like dark, damp places and feed on tiny plants in the soil.

Woodlouse: A small, usually gray animal with a hard, armored skin. It is a **crustacean** and lives in dark, damp places.

Earwig: A small, brown, shiny insect with pincers at the back of its body. It usually lives under stones or in cracks in the wall or ground.

Glossary

Antennas Feelers on an insect's head used for smell, taste and touch.

Arachnid An animal with eight legs, like a spider, harvestman, or mite.

Backbone The line of bones down the middle of the back.

Burrow An underground animal home made by digging out soil.

Cell In a bees' or wasps' nest, the six-sided (hexagonal) space used for an egg, larva, or food.

Colony A large group of insects that live together, such as ants or wasps.

Crustaceans A large group of animals, usually with a hard skin or shell, which includes woodlice, crabs, and lobsters.

Fertile Able to support a lot of plant growth.

Grub The name given to a beetle larva.

Habitat A place where plants and animals live.

Hibernation Many animals survive cold weather by going into a deep sleep, called hibernation.

Hive A place where bees build honeycombs.

Insects A huge group of animals. All insects have a body in three parts—the head at one end, thorax in the middle, and abdomen at the end. Six legs are attached to the body and many insects have two pairs of wings.

Invertebrates A huge group of animals without a backbone, including insects, worms, and spiders.

Larva (plural, larvae) The stage in the life cycle of many insects after they hatch from eggs.

Leaf litter A pile of dead leaves, pieces of bark, and other plant parts.

Mollusks A large group of animals with soft bodies. It includes slugs, snails, mussels, octopuses, and oysters.

Nectar The sweet liquid made by flowers and eaten by many insects.

Nocturnal Animals that rest during the day and are active at night.

Pollen A fine powder made by the flower of a plant.

Predators Animals that eat other animals, rather than plants or dead things.

Queen In wasps', ants,' and bees' nests, the most important female of the group who lays all the eggs.

Recycle To put through a process that allows used things to be reused.

Rot When living things die, break down, and decay.

Saliva Watery liquid in the mouth of an animal.

Segment A section of something. Some minibeasts have segmented bodies, meaning they are made up of many similar sections joined together.

Solitary Living alone, like the solitary bee.

Web Sites to Visit

http://dev.pestworldforkids.org/guide.html
This web site is packed with fun facts and photos about minibeasts. Learning games and a "Pests Quiz" are also featured!

http://kidshealth.org/kid/ill_injure/index.html
Check out this web site for helpful advice on how to take care of your bug bite.

Note to Parents and Teachers:
Every effort has been made by the publishers to ensure that these web sites are suitable for children, that they are of the highest educational value, and that they contain no inappropriate or offensive material. However, because of the nature of the Internet, it is impossible to guarantee that the contents of these sites will not be altered. We strongly advise that Internet access is supervised by a responsible adult.

Index